SCIENCE MUSEUM

# X-RAYS

*their discovery and applications*

BRIAN BOWERS, B.SC.(ENG.), A.K.C., C.ENG., M.I.E.E.

*LONDON*
HER MAJESTY'S STATIONERY OFFICE
1970

1   Equipment constructed by Dr Russell Reynolds in 1896–7
and used in his medical practice

# Contents

# Introduction

X-rays are produced whenever fast moving electrons hit a solid object.

Towards the end of the nineteenth century many physicists were taking an interest in 'cathode rays' – streams of electrons emitted from the cathode (negative electrode) when a high voltage was applied between two electrodes in an evacuated tube. On the evening of 8 November 1895 Professor Röntgen, Director of the Physical Institute of the University of Würzburg, was studying cathode rays in his laboratory. His glass discharge tube was enclosed in black card to mask the fluorescent glow which always accompanies such experiments. While the tube was working he noticed a glow in some fluorescent material nearby. He investigated and found that the fluorescence was caused by some previously unknown agent generated within the tube.

He called this agent 'X-rays'.

This booklet gives an account of Röntgen's discovery and describes the development of the use of X-rays for both medical and industrial purposes. The author is grateful to Miss R.J. Plesters of the National Gallery for advice on the use of X-rays in the study of paintings, and to Dr Keith James M.A., M.B., F.F.R. for assistance with the section on medical applications. The photographs of paintings are reproduced by courtesy of the Trustees of the National Gallery, London.

# The X–ray collection in the Science Museum

The Science Museum possesses an extensive collection of X-ray tubes, including examples of all the types mentioned in this booklet. There are also fluorescent screens formerly belonging to some of the pioneers of X-radiography and a number of early radiographs.

For the general visitor the greatest interest is probably to be found in the early X-ray installations. The equipment (figure 1) made by Dr Russell Reynolds in 1896 and 1897 was subsequently used by him in his medical practice in London. The power source is a battery of six cells of the simplest construction complete with a windlass for raising the electrodes from out of the cells when the battery is not in use. The high voltage supply needed for the X-ray tube is generated by an induction coil, which is supported on glass insulating pillars on a low table and supplied through a motor-driven mercury contact breaker. The wooden box under the table houses a capacitor and the X-ray tube itself is mounted on a wooden stand.

The bare high-voltage wires connecting the X-ray tube exposed the patient to a very real risk of electric shock. This was particularly the case in dental equipment where it is essential to have the patient close to the tube. The Collection includes one early dental X-ray set fitted with an earthed metal grid to keep the patient away from the 'live' end of the tube. As X-ray equipment developed, the high voltage wires and the tube were enclosed in an earthed metal screen, and today small X-ray tubes such as are used in dentistry are combined with their supply transformer in a single metal enclosure.

The pioneers of X-rays, such as Russell Reynolds, assembled their installations from whatever components they could make or obtain. Early commercial X-ray installations were also a collection of separate parts assembled together for the purpose. This may be contrasted with modern installations which are designed as a whole, with the high voltage supply, control equipment and tube all being integral parts of the installation rather than separate units connected together.

An early commercial X-ray equipment (figure 2) made by Butt & Co at about the end of the First World War has a wooden cabinet housing the induction coil and rotary contact breaker. The control equipment is mounted on one side and the top of the cabinet supports an ammeter and a spark gap. The spark gap serves as a voltmeter for the high voltage supply to the tube (the tube is not shown, but would be connected by bare wires to either side of the spark gap).

Among the more modern items in the Collection are parts of a betatron formerly used for X-ray therapy at the Christie Hospital and Holt Radium Institute in Manchester. This accelerated electrons in a circular path about 18 inches in diameter until they reached a velocity corresponding to 20 million volts, when they were allowed to strike the target and yield extremely penetrating X-rays. There are also modern X-ray tubes including a rotating anode diagnostic tube of the type now used in medical radiography. Some current applications of X-rays are featured in photographic displays.

2 Equipment made by Butt & Co about 1919

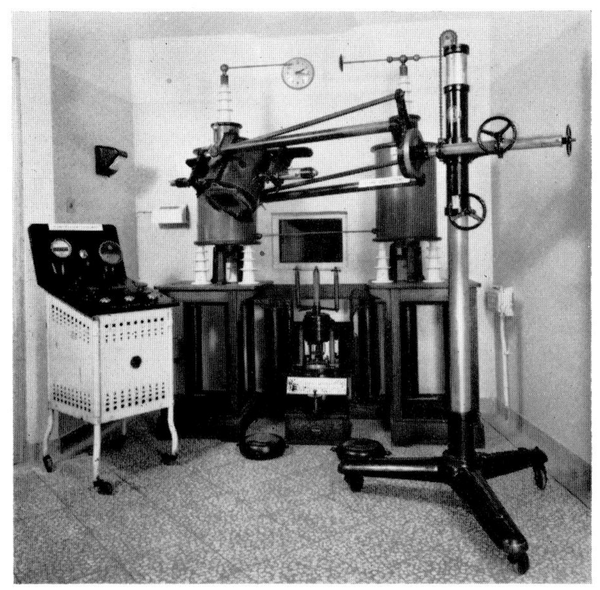

3 200 kV X-ray therapy equipment installed at Guy's Hospital, London, in 1923

# The discovery of X–rays

## The background to the discovery

Two great advances in basic physics became known during the closing years of the nineteenth century. One was the discovery of X-rays; the other was the understanding of the electron. Both arose from the study of electric discharges in gases at low pressure. If two metal electrodes are sealed into a glass tube containing air at atmospheric pressure and then connected to a source of electricity nothing is apparent unless the voltage is very high – about 30,000 volts for every centimetre between the electrodes. If the pressure is reduced an electric current flows through the gas in the tube and a variety of phenomena occur at much lower voltages. At a pressure of about ten millibars a steady glow fills the space between the electrodes. If the pressure is further reduced the glow breaks up into bands of alternate light and dark. At a pressure of about 0·2 millibars the glow disappears altogether, although a current is still flowing through the tube, but the glass of the tube begins to fluoresce brightly.

The German physicist Plücker concluded in 1859 that the fluorescence was due to something radiating from the negative electrode, or cathode. These 'cathode rays' were closely studied during the remainder of the nineteenth century. The English physicist Crookes published the results of a series of researches on the subject in 1879.

He showed that the cathode rays were emitted normally to the cathode surface and could be deflected by a magnet. When he focused the rays to a point sufficient heat was developed to melt glass or platinum foil. He concluded rightly that the rays consisted of negatively charged particles, but although his experimental results were indisputable the conclusions he drew from them were vigorously contested. Most German physicists thought that the cathode rays were a wave motion similar to light and that there was no propagation of matter.

Hertz, working at Bonn, was seeking an experimental proof of Maxwell's theories of the nature of electricity and magnetism and he took up the study of discharges in a vacuum. He found that the cathode rays would pass through a thin film of gold or aluminium placed in its path. After his death in 1894 his pupil Lenard continued his work. Lenard made a tube with a thin aluminium window and succeeded in bringing cathode rays into the outside air. He found that they still produced fluorescence, but that the rays would not travel far through air at atmospheric pressure. Lenard said that the cathode rays passed through the hand. This was almost certainly an observation of X-rays produced where the cathode rays struck the window of this discharge tube, but he failed to notice that it was a different kind of ray. Sir William Crookes often made the observation that photographic plates which happened to be stored near his tubes became fogged and this must have been caused by X-rays. On one occasion he returned some plates to the manufacturer as unsatisfactory! Many workers must have produced X-rays accidentally while studying cathode rays, but only Röntgen observed their presence, and realised that he was dealing with a 'new kind of rays'.

4  W.C. Röntgen (1845–1923)

## W.C. Röntgen

Wilhelm Conrad Röntgen was born on 27 March 1845 in Lennep in the German Rhineland. He was the only child of a cloth merchant and manufacturer, and may have been a distant relative of David Röntgen (1743–1807) who had achieved fame as a cabinet maker. When he was three years old his parents moved to Apeldoorn in Holland, the home of his mother's parents, and the family became Dutch instead of Prussian citizens.

Röntgen went to a boarding school in Apeldoorn and later to Technical School in Utrecht. He was expelled from the Utrecht school after refusing to name the fellow student who drew a caricature of the teacher. At school he was not an exceptional pupil, though he showed an aptitude for mechanical things. He was a lover of nature and all his life liked to spend his holidays in the Alps or among the Lakes of North Italy. In later life he preferred travelling in a horse drawn carriage to using a motor car.

Early in 1865 Röntgen attended the University of Utrecht, though not as a regular student since he lacked the required qualifications. He discovered that it was possible to enter the Zurich Polytechnical School in Switzerland by passing an entrance examination and without school references. He passed the entrance examination and in November 1865 he became a student of mechanical engineering in Zurich, and the stigma of his expulsion from school was left behind. For the rest of his life Röntgen retained happy memories of his year in Zurich and a sense of gratitude to Augustus Kundt, the professor of physics at the Polytechnical School, who inspired Röntgen to follow a career in physics. Röntgen became Kundt's assistant at Zurich and then at the University of Würzburg when Kundt took a post there.

While in Zurich Röntgen met Anna Bertha Ludwig, whom he married in 1872. They had no children of their own, but adopted Anna's niece.

Despite all Kundt's efforts on his behalf Röntgen could not get an academic post at Würzburg. In 1872 Kundt went to the newly founded University of Strasbourg (then in Germany) taking Röntgen with him. Two years later Röntgen became a 'Privat Dozent' at Strasbourg and his parents moved there from Apeldoorn to be near their son. In 1875 he succeeded H.F. Weber, on Weber's recommendation, as Professor of Physics and Mathematics at the Hohenheim Agricultural Academy in Würtemberg. He was not happy at Hohenheim and returned to Strasbourg a year later as Professor of Theoretical Physics. From 1879 to 1888 he held the chair of Physics at Giessen University, and because of his excellent work during these years was offered posts at Jena and Utrecht, which he declined. However, in 1888 he accepted an invitation to become Professor of Physics and Director of the Physical Institute of the University of Würzburg. Thus he became a head of department in the University which had earlier refused him any academic position.

At Würzburg he turned his attention – as many scientists were then doing – to the study of cathode rays. It was his custom when taking up any new investigation to repeat experiments made previously by others working in the same field. For one of these experiments he had a Crookes tube covered with black card to mask the fluorescent glow which always exists in the glass in this kind of experiment. When the tube was connected up he noticed that some crystals of barium platino-cyanide which happened to be on a table nearby became fluorescent. The observation was made on the evening of Friday 8 November 1895 'at a late hour when assistants were no longer to be found in the laboratory'. Röntgen investigated and quickly satisfied himself that the tube was emitting some hitherto unknown kind of ray which produced the fluorescence. A screen coated with barium platino-cyanide and held near the tube fluoresced all over, but if a metallic object was placed

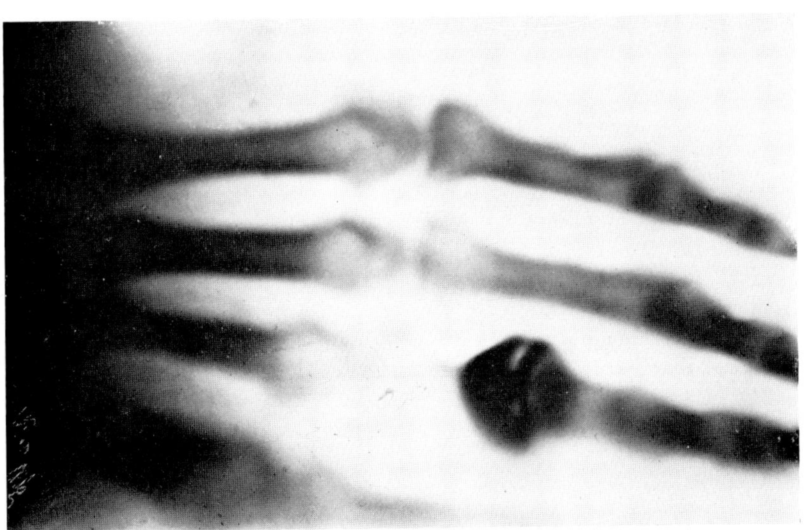

5   Röntgen's hand, X-rayed in December 1895

between the tube and screen it cast a shadow. Röntgen told his friend Boveri, 'I have discovered something interesting but I do not know whether or not my observations are correct', but with the exception of this one remark he told no one of his discovery for seven weeks. During that period he devoted himself to the study of the 'something interesting', and prepared a paper setting out the main properties of X-rays.

Röntgen himself used the term X-rays ('X-strahlen' in German), because the nature of the rays was uncertain. In Germany the name Röntgen-rays was adopted almost immediately and remains in use today. The discoverer's name has rarely been used in English-speaking countries, probably because the English speaker does not know how to pronounce 'Röntgen'. The word 'Skiagraph' (from the Greek for a shadow) was introduced in 1896 for a picture obtained by X-rays and was used for a few years. 'Radiograph' was also introduced in 1896 with the same meaning, and in this country nowadays a picture obtained by means of Röntgen's rays is invariably called either an 'X-ray' or a 'radiograph'.

## The announcement of Röntgen's discovery

Röntgen's first paper on X-rays was handed to the president of the Physical Medical Society of Würzburg on 28 December 1895. Before that day no one – not even his closest assistants – had been told of the discovery. The paper, which Röntgen called a 'preliminary communication' is remarkable for the wealth of detail it contains. After announcing his basic discovery that a new kind of ray is produced in a working discharge tube, and that this ray passes through glass, black card and at least two metres of air, he continues with a study of the relative transparency to the rays of many different substances. He concludes that the transparency varies with the density of the substance. He reports that he cannot obtain either reflection or refraction, from which it follows that a beam of X-rays cannot be concentrated by means of mirrors or lenses. He notes that X-rays originate at the point where the cathode ray beam strikes the glass, and radiate in all directions, and that X-rays are also produced if the cathode rays strike an aluminium insert instead of the glass. The nature of X-rays puzzled him. They were not cathode rays from the main discharge which had bounced off the glass because they were not deflected by a magnet. The rays could not, he thought, be ultra-violet light, because they were not reflected or refracted as ultra-violet light was. The new rays must be related to light, argued Röntgen, because they travel in straight lines, cause fluorescence and act on a photographic plate. He adopted the theory, later rejected, that the new rays were *longitudinal* vibrations in the ether, in contrast to the *transverse* vibrations of infra-red, visible and ultra-violet lights.

In the paper Röntgen listed a number of 'shadow-pictures' he had photographed. One was of a door to the room containing the X-ray equipment, the photographic plate being outside. Others showed a set of weights in a closed box, and a piece of metal whose inhomogeneity was revealed by the X-rays. But the picture which most captured the public imagination was of the bones in a living hand.

Röntgen's paper was immediately printed and distributed. An English translation was published in *Nature* on 23 January 1896 and within a few weeks the news of the discovery had spread throughout the world.

# The early days of X–rays

## The popular response to Röntgen's discovery
The English physicist Sir J.J. Thomson remarked that Röntgen's discovery
appealed to the strongest of all human attributes – curiosity! The first newspaper report was the result of a 'leak'. Very soon after the discovery Röntgen
had sent a few of his first X-ray pictures to Franz Exner at Vienna University.
Exner, who had worked with Röntgen in Zurich and Strasbourg, showed the
pictures to a number of his colleagues. One of these was the son of the editor
of the Vienna *Presse*. He showed some of the photographs to his father, who
immediately realised the news value of the discovery and published an article
setting out the main facts of Röntgen's discovery. The *Presse* report, which
included a prediction of the possible medical value of the rays, was quickly
copied by papers in many different countries, including the London *Standard*
of 8 January 1896.

6   Cartoon from *Punch*
7 March 1896

The properties, real and imagined, of the new rays suggested new ideas to
the contemporary humorists. The following examples are from periodicals of
April and May 1896:

*Dealer* (pointing towards a thin horse): 'There, sir! That's what I call a
picture!'
*Prospective Buyer:* 'H'm, yes, he does rather suggest one of those roentgen
photographs!'

*Guest* (to waiter who has brought him chops with a large bone and little
meat): 'Waiter, is this a chop à la Röntgen?'

*She:* 'I wish some photographs taken.'
*Photographer:* 'Yes, Madame, with or without?'
*She:* 'With or without what?'
*Photographer:* 'The bones.'

The *Punch* cartoon 'The March of Science' is of interest, apart from the humour, for the light it sheds on popular understanding of X-rays at the time. The cartoonist makes play of the ability to see through a wooden door, but he shows the figure behind as though viewed normally rather than as a skeleton. Another common misunderstanding was to imagine the X-ray source located with the camera and the picture taken by reflected 'light': whereas, of course, X-ray pictures are invariably taken by transmitted rays with the object placed between the X-ray source and the photographic film. This is seen in the *Life* cartoon 'Look Pleasant, Please'.

During the months following Röntgen's discovery, the new rays were credited with the powers of the 'philosopher's stone', and brought into the discussion of such subjects as vivisection, the temperance and prohibition movements, spiritualism, fortune-telling and telepathy. In Iowa a farmer who had been experimenting with X-rays was reported to have transformed a cheap piece of metal worth 13 cents into $153 worth of gold by means of X-rays. The gold had been tested and found pure! On the subject of vivi-

THE NEW ROENTGEN PHOTOGRAPHY.
" LOOK PLEASANT, PLEASE."

7   Cartoon from *Life* 27 February 1896

section, *Life* observed: 'If the Röntgen method of seeing through things pans out anywhere near as well as its friends expect, we are entitled to hope that it will almost put an end to vivisection. There will be no need to put a knife into a live animal when a ray will make its inner workings visible'. A leading temperance worker, Miss F.E.Willard, remarked optimistically: 'I believe the X-rays are going to do much for the temperance cause. By this means drunkards and cigarette smokers can be shown the steady deterioration in their systems, which follows the practice, and seeing is believing'.

One of the more fantastic reports was the announcement in a New York newspaper that the College of Physicians and Surgeons were using X-rays 'to reflect anatomic diagrams directly into the brains of the students, making a much more enduring impression than the ordinary methods of learning anatomic details'.

## The scientific response to Röntgen's discovery

The enthusiastic reception given to Röntgen's discovery by the popular press was matched by the scientific world. In 1896 alone about fifty books and pamphlets and almost one thousand papers were published on the subject. Medical, general scientific and photographic journals all published articles on the discovery and its potential applications. The first journal devoted solely to X-rays made its appearance in May 1896 under the title *Archives of Clinical Skiagraphy.* This was founded by Sidney Rowland and published in London. The following year the name was changed to *Archives of the Roentgen Ray,* and similar journals appeared in other countries.

Few discoveries – if any – have captured the imagination and been announced around the world in so short a period of time as was the discovery of X-rays. But the impact of the news on the scientific world was enormously increased by the fact that every physical laboratory possessed the equipment necessary for generating the new rays. After the researches of Crookes and others, vacuum tubes and induction coils were standard equipment, and little more was needed to confirm the truth of Röntgen's results and to investigate further the properties of the new rays.

In May 1896 the *American Electrician* began a three part series on how to make 'A Röntgen Ray Outfit'. This described how to make an induction coil capable of producing a three-inch spark and how to make a rotary contact breaker. But it advised purchasing a Crookes' tube and recommended one with thin glass, and of German manufacture (because, it said, German glass contains less lead than American glass).

A catalogue of 'Röntgen Ray Apparatus' was issued by the American General Electric Company in the autumn of 1896. The manufacture of X-ray equipment was then established on a commercial scale, but it was to be several years before a user could purchase a complete X-ray equipment in one unit, rather than a number of separate items – induction coil, contact breaker, tube, etc. – which had to be assembled and interconnected.

One of the first to experiment with X-rays in America was T.A. Edison. For medical examinations he advocated the use of fluorescent screens rather than photographic plates so that the doctor could see fractures etc. immediately without having to wait for the plate to be developed. After testing every chemical available to him (nearly two thousand different substances) he decided that the material which fluoresced best under the influence of

X-rays was calcium tungstate, which gave an image several times brighter than the barium platino-cyanide used by Röntgen.

## Röntgen's subsequent work on X-rays

Röntgen demonstrated his discovery to Kaiser Wilhelm II, who was always interested in scientific developments, in Berlin on 13 January and then lectured on the new rays to the Physical Medical Society of Würzburg on 23 January 1896, less than one month after his first announcement. This was probably his only lecture on the discovery to a large audience. Röntgen related how he had made his discovery and showed a number of X-ray pictures. He seems to have had some reluctance about speaking on his work at that time when his experiments were still being developed, but he did so because of the great public interest that had been aroused. During the meeting Röntgen took an X-ray photograph of the hand of the anatomist A. von Kölliker, who was in the audience. After this had been done, von Kölliker proposed that henceforth the new rays should be called 'Röntgen's rays', and this suggestion was approved with great enthusiasm by the audience.

Röntgen submitted a second paper on X-rays to the Würzburg Physical Medical Society in March 1896. Most of this paper is an account of experiments showing that air (or any other gas) which has been exposed to X-rays could conduct electricity and would discharge an electrically charged body. He could not explain the phenomena he observed, but when J.J. Thomson's work on the electron became known it was possible to explain Röntgen's observations by saying that the radiation 'ionised' the air (that is, stripped some electrons from the atoms) making it conductive. The ionising property of X-rays is exploited in some procedures for measuring the rays.

In the same paper Röntgen says that it can be advantageous to include a Tesla apparatus in the circuit between the induction coil and the X-ray tube. More intense X-rays are obtained, and some tubes which are either too hard or too soft to be operated directly from an induction coil will work satisfactorily through the Tesla apparatus. This was an oscillatory circuit which causes the voltage across the X-ray tube to reach a higher peak value than would otherwise be the case. The Tesla principle was often exploited in early X-ray equipment, since then the induction coils did not have to withstand so great a voltage, and has occasionally been used in more recent equipment.

His third and final paper was published in May 1897, and contains an account of several further experiments and measurements. Röntgen found that any substance – even air – which was subjected to X-rays would itself emit X-rays. He showed that this phenomenon was truly the generation of fresh X-rays ('secondary radiation') and not mere scattering of the main beam. With a device similar to an optical photometer he compared the X-ray outputs of different tubes and also examined the distribution of the output from a single tube. He found that with soft X-rays the emission was uniform over a hemisphere centred on the target, though with hard X-rays and a thin target some radiation appeared from the back of the target. Finally he made a number of comparative measurements of the opacity to X-rays of different thicknesses of various substances.

In his three papers Röntgen described most of the properties of the rays he had discovered, but the actual nature of the rays remained unknown at that time. The later history of X-rays is chiefly the story of technical improvements in the equipment and the refinement of techniques.

Röntgen was very reticent by nature and the many honours bestowed on him were more of a burden than a pleasure. The University of Würzburg gave him the honorary degree of Doctor of Medicine, and he accepted honorary citizenship of his native Lennep. He declined nearly all invitations to address scientific societies on his discovery. He declined the offer of nobility by the Prince Regent of Bavaria, who later bestowed the title 'Excellency' on Röntgen. In 1901 he became the first Nobel laureate for physics, and travelled to Stockholm to receive his prize though he did not give a Nobel Lecture.

In 1900 Röntgen left Würzburg to take charge of the Physical Institute of the University of Munich, where he resumed his earlier work on the physical properties of crystals. After his retirement in 1920 he was given permission to use two rooms in the Institute. He continued to work there until a few days before his death on 10 February 1923.

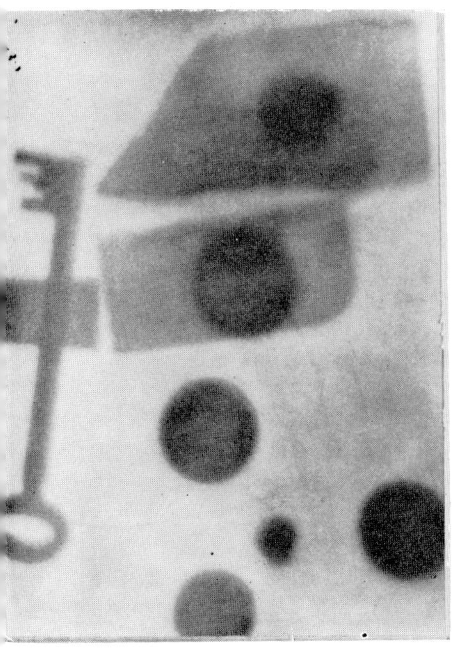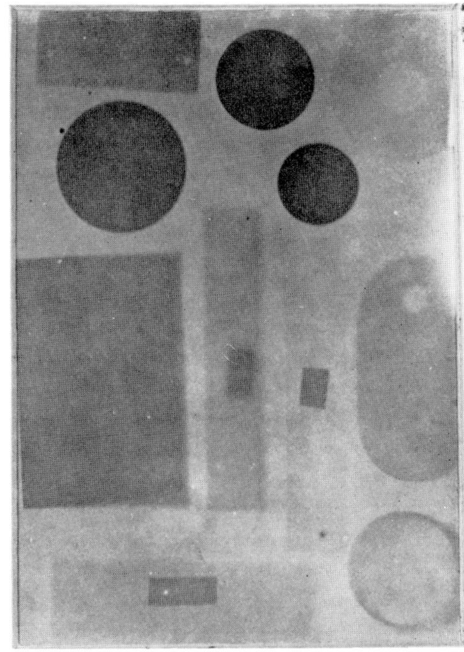

8   X-radiographs taken by A.A. Campbell Swinton in London on 8 January 1896

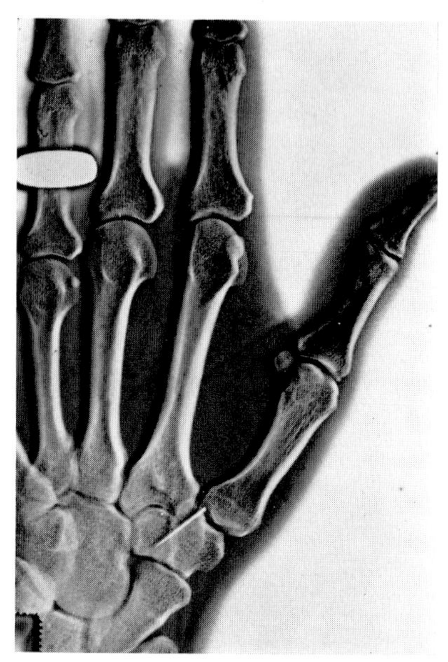

9 Early radiograph showing a piece of a needle in a hand. This is an example of C. Thurston Holland's 'plastic prints' made by preparing a positive from the original negative and then printing through the original negative and the positive superimposed.

10   Edison's Surgeon's X-ray Apparatus 1896

# X–ray tubes

## Early X-ray tubes

The essential components of an X-ray tube are an airtight vessel, usually of glass, and two electrodes sealed into it. In the early tubes the vessel is evacuated to a low pressure, but some gas molecules remain (for which reason these are known as 'Gas-Tubes'). If such a tube were completely evacuated it would not function. The electric discharge when a high voltage is applied between the electrodes causes ionisation of the gas atoms, and the positive ions are driven towards the cathode (negative electrode) by the electric potential across the tube. This bombardment of the cathode by the positive ions causes the emission of electrons which, on striking the target, generate X-rays. The target – sometimes called the anti-cathode – may be the wall of the vessel, the anode electrode or a separate metal insert connected to the anode.

The tube with which Röntgen discovered X-rays was probably a pear shaped 'Crookes' tube' with the cathode at the pointed end and the anode to one side. The X-rays were generated in the area where the cathode rays struck the glass at the larger end of the tube. To get a sharp shadow picture the X-ray source must be as small as possible, and for the earliest X-ray pictures the end of the tube was covered with a lead sheet having a small hole in it. Only the rays passing through the hole could be utilised, so that the system was extremely inefficient and long exposure times were required.

The 'Focus' tube designed by Professor Herbert Jackson was designed to focus the electron beam on to a small target area in order to produce as small an X-ray source as possible. He achieved this by using a concave cathode. Since the electrons are emitted perpendicularly to the cathode surface and then travel in virtually straight lines (irrespective of the position of the anode) they converge at the centre of curvature of the cathode, and the target is

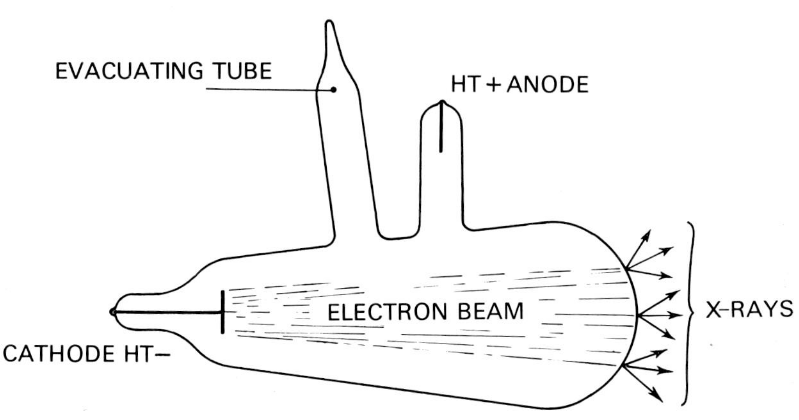


11   Production of X-rays from a 'Crookes' Tube.

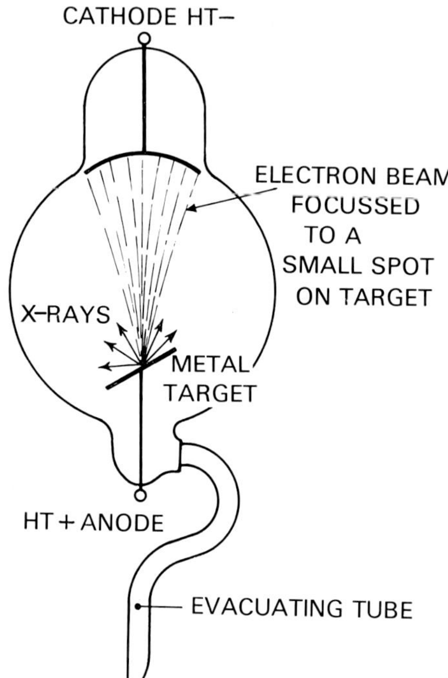


12 Focus Tube. The concave cathode focuses the electron beam on to a point on the surface of the target, which is now a piece of metal connected to the anode instead of the glass of the tube.

placed at this point. Crookes had used tubes with cathodes shaped in this way in the course of his studies of cathode rays, but Jackson seems to have been the first person to suggest using the idea in X-ray tubes, which he did in March 1896.

In his first paper on X-rays Röntgen noted that X-rays were also generated if the cathode rays struck an aluminium insert instead of the glass of the tube. In the second paper, published in March 1896, he reported that every substance which he had tried would yield X-rays when used as a target and bombarded by cathode rays in a discharge tube. He found that there were qualitative differences in the X-rays generated using targets of different materials and that he could produce the most penetrating X-rays with a tube having a concave aluminium cathode and a platinum anode as a target inclined at 45° to the axis of the cathode.

Many scientists in different countries experimented with the new rays during 1896. Tubes of different shapes and targets of various metals were tried. By the end of that year it was established that the shape of the tube did not matter, and that the best target metals were those of highest atomic weight. Tungsten (atomic weight 184) and uranium (at. wt. 238) were used experimentally, though platinum (at. wt. 195) was preferred because it is easier to work. Aluminium (at. wt. 27) was sometimes used, despite its low atomic weight, because it remains stable in a vacuum discharge whereas a tube with a platinum target tends to become coated inside with a thin layer of platinum black produced by the destructive effect of the discharge on the metal.

The first X-ray tubes were of very low power by modern standards and exposures of long duration – at least several minutes – were required for most purposes. The penetrating power of the X-rays and the speed with which an adequate photograph could be obtained were found to increase as the voltage across the X-ray tube was increased. However, most of the energy in the electron beam is converted into heat where the beam strikes the target, and as increasing power was applied to X-ray tubes it became necessary to increase the mass of the target to prevent overheating. (The thin platinum targets used in some of the early tubes could easily be vapourised by the electron beam.) A large, solid, platinum target would be very expensive, and platinum plated nickel was adopted. This construction remained in general use until the introduction of tungsten, which is still the preferred metal for the target. Even with a tungsten target, however, the ability of the target to withstand the effect of the electron bombardment is a limiting factor in the rating of the tube. The working surface of the target develops fine cracks, with the result that a proportion of the electron beam generates X-rays at the sides or bottom of the cracks. The metal of the target screens such radiation and the useful radiation from the tube therefore falls as the surface of the target loses its initial smooth finish.

In early tubes the target was attached to a massive metal block which prevented excessive rise in temperature by absorbing the heat generated. Then the metal block was brought through the glass wall of the tube and attached to external radiating fins. In some tubes removable cooling tongs could be inserted in the block from outside the tube, and as soon as the target and tongs became too hot the operator could remove the tongs and replace them with a cool pair.

Cooling water was circulated in a hollow target in some tubes. The arrangement was effective, but suffered from the disadvantage that the cooling water system including its external radiator had to be at the voltage of the target. A

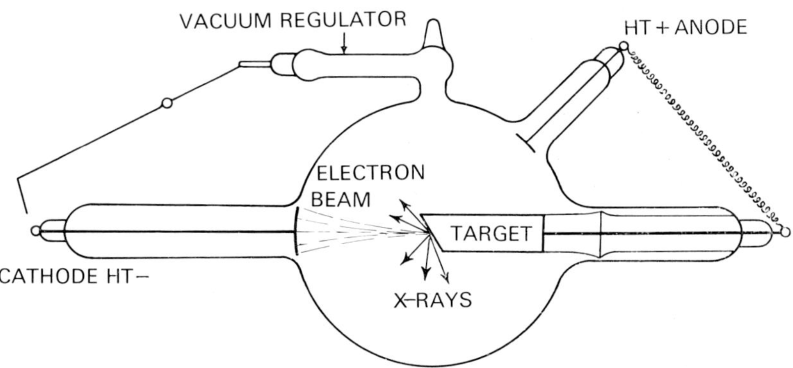


13   Typical early X-ray Tube with vacuum regulation

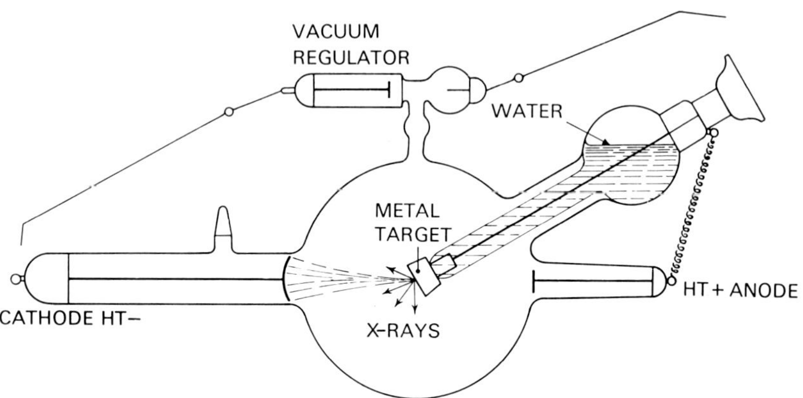


14  Boiling Water X-ray Tube. The temperature of the target cannot rise above the boiling point of water and the tube operates under steady conditions for long periods.

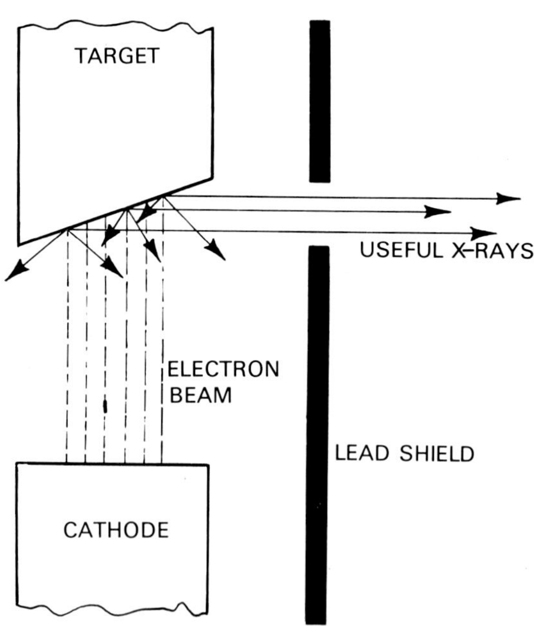


15  Principle of Line Focus X-ray Tube. The electron beam is ribbon-shaped and strikes the target along a narrow strip, instead of being focused on to a point (hence the term 'line focus'). Because of the foreshortening effect of the angle of the target surface the 'Useful X-rays' appear to come from a source much smaller than the true area of the strip. Only the 'Useful X-rays' can pass through a window in the lead shield.

more practical arrangement was the boiling water tube, in which the target assembly and a reservoir above it contain water which boils steadily while the tube is operating. The temperature of the target remains steady and no connecting water pipes are required.

The need for a small focal spot on the target means that intense heat is developed in a very small area and has to be conducted away through the target metal. The ability of the metal to conduct heat proved to be the factor which limited the power which could safely be applied without damaging the target, and hence limited the intensity of X-rays which could be produced. The 'line-focus' tube was devised to alleviate this situation. In this type of tube the working area of the target is a narrow strip rather than a spot and the cathode is designed to emit a ribbon-like beam of electrons. The target strip is inclined at such an angle that, viewed in the direction in which the X-ray output is taken, the strip is fore-shortened to a square of side equal to the width of the strip. The X-rays appear to come from a small square source, although the much greater area of the target strip is available to conduct away the heat generated.

A further development is the rotating target tube in which the working portion of the target is continuously changing so that a still greater area of metal is effectively available to conduct away the heat. The line-focus principle is used, but the target strip is a continuously changing portion of the sloping side of a disc rotated at high speed by an induction motor. Most modern X-ray tubes use this principle.

Until the 1920s most X-ray tubes were 'gas' tubes, which depend on the presence of some residual gas in the tube for the electric discharge to take place. In operation some gas molecules become adsorbed in the glass of the tube, so that the vacuum tends to increase and the tube becomes 'harder', requiring a higher operating voltage. It was recognised that in order to secure satisfactory and uniform working of these gas tubes the degree of vacuum had to be kept approximately constant. Hence numerous regulators were devised which controlled the vacuum by releasing small quantities of gas into the tube.

The metal palladium has the property that when heated to incandescence hydrogen can pass through it. One system of vacuum regulation for X-ray tubes devised by Professor Villard in Paris in 1898 made use of this property. A palladium tube was sealed into the side of the glass bulb. To 'soften' the X-ray tube the palladium tube was heated by a gas flame. Hydrogen in the flame then passed through the palladium into the bulb, so reducing the vacuum.

Other regulators employed substances which absorb gases, such as charcoal or mica. A small quantity of one of these substances was placed in a glass side tube which could be heated, when some of the absorbed gas would be driven off.

An automatic vacuum regulating device was introduced about 1900. Mica containing absorbed gas was placed in a side tube and could be heated by a discharge between auxiliary electrodes connected in series with an external spark gap. As the tube hardened the supply voltage rose. When the supply voltage became sufficient to produce a spark across the external spark gap a discharge took place between the auxiliary electrodes. This heated the mica and gas was driven off to reduce the vacuum and soften the tube.

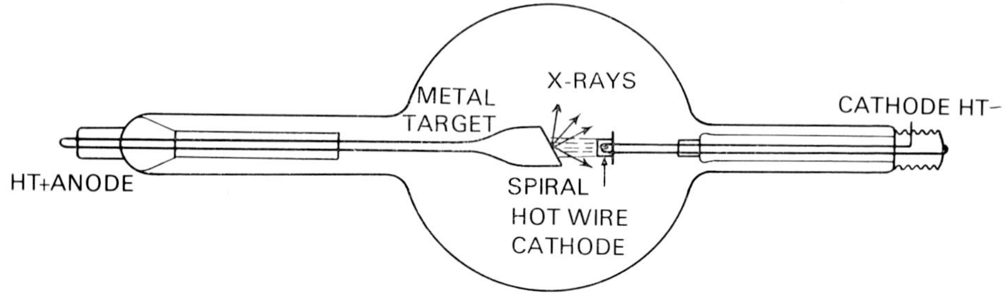



16   The thermionic X-ray tube

## Thermionic X-ray tubes

The X-ray tubes described so far are all 'gas' tubes, which depend for their operation on the presence of some residual gas in the tube. One disadvantage of the 'gas' tube – the variation of the degree of vacuum over a period of time – has already been mentioned. Another disadvantage is that the voltage across the tube and the current through it are inter-dependent. The reason this matters is that the *penetration* of the X-rays depends on the voltage, and the *intensity* depends on the current. (The *penetration* is a measure of the distance the X-rays can pass through a body being radiographed; the *intensity* is a measure of the degree of blackening obtained on the photographic plate in a given time.) In the gas tube a reduction in gas pressure (a 'harder' tube) leads to a higher operating voltage and hence more penetrating X-rays. But, because there is less gas in the tube, the current falls and the intensity of the X-ray beam is reduced.

In the thermionic X-ray tube, devised by W.D. Coolidge in the USA in 1913, the vacuum is as near perfect as possible, and the electrons are produced by 'thermionic emission'. The cathode is no longer a cold metal block, but a tungsten filament heated to incandescence by a current passing through it. This filament is usually mounted within, and electrically connected to, a shaped metal block which focuses the electron beam on to the target. The intensity of the X-ray output depends only on the current in the electron beam, which is a function of the temperature of the filament, and this can be controlled by adjusting the current in the filament. The penetration of the X-rays produced depends – as with the gas tube – on the voltage between cathode and target. With this arrangement the intensity and penetration of the X-ray output are independently controlled by adjustment of the filament current and tube voltage, respectively.

All modern X-ray tubes are of the thermionic type.

Once it was realised that excessive exposure to X-rays was dangerous, steps were taken to screen the tube so that 'stray' radiation was not emitted. Some early, low power, tubes are made mostly of lead glass, which is comparatively opaque to X-rays, but with a small 'window' of lime glass through which the X-ray beam emerges. In this way all radiation except the main

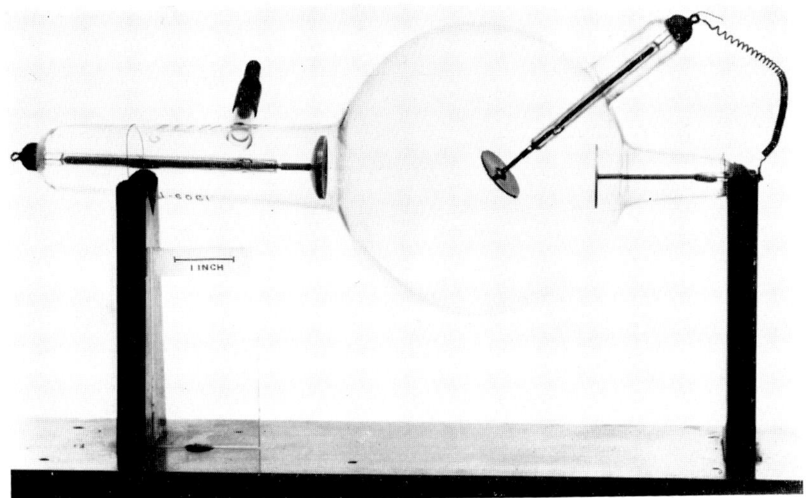

17   Focus Tube. The concave cathode focuses the electron beam on to the target, which is a flat metal disc in the centre of the tube.

18   Tube with cooling tongs. The target is hollow and the cooling fins are mounted on a split copper rod which pushes into the target. When the tube becomes hot, the tongs are replaced by a cool pair

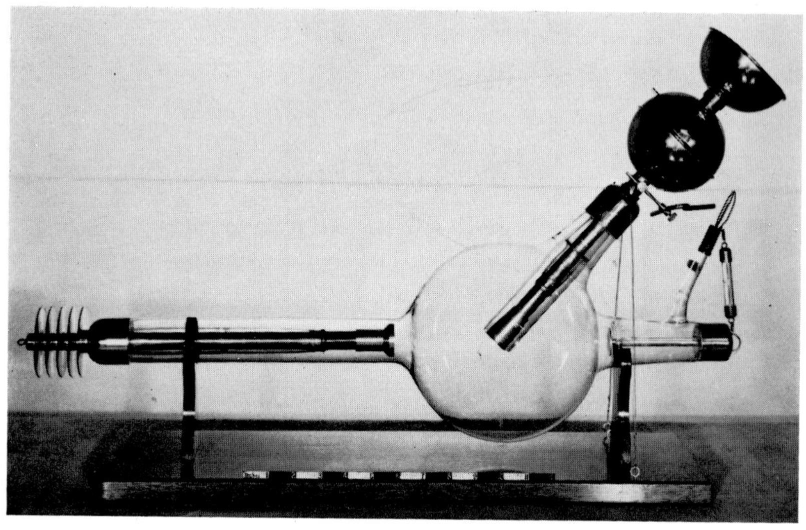

19   Boiling Water Tube. The glass tube around the target and the copper sphere above it contain water. The cathode (left) also has cooling fins.

beam is screened from the operator or patient. In the 1920s an entirely new design of X-ray tube – the 'Metalix' tube – was developed in which the X-rays are generated inside a metal chamber forming the main part of the tube. This screens all radiation from the tube except the main X-ray beam which passes through a glass window. The metal is an alloy of chromium and iron which can be sealed directly on to the glass endpieces to form a good vacuum seal. Modern X-ray tubes are usually enclosed in an earthed metal housing which protects the user from electric shock and incorporates a layer of lead to prevent the emission of unwanted radiation.

## Accelerators of other designs
X-ray tubes and equipment working as described above have been operated at up to a quarter of a million volts, but the practical difficulties are immense and for still more penetrating radiation other techniques are adopted.

All X-ray tubes are 'accelerators', devices which accelerate electrons to a high speed before they strike the target. The penetrating power of the X-rays generated depends on the speed with which the electrons strike the target; this speed is determined solely by the voltage across the tube. For some applications, such as the radiographic examination of thick metal structures, X-rays of very high penetration are required.

By using machines such as the linear accelerator or the betatron which have been developed for research into atomic physics, electrons may be accelerated to speeds which would require several million volts across a conventional X-ray tube. In these machines electrons are repeatedly accelerated across the same potential drop so that a voltage multiplying effect is obtained before they strike the target.

In many applications, both in medicine and industry, very high energy X-rays are being replaced by gamma rays from radio-active isotopes. Gamma rays occupy the part of the electro-magnetic spectrum adjacent to X-rays and may be regarded as very hard X-rays. The isotopes used, such as cobalt 60, need careful shielding, but this may be less of a problem than the elaborate equipment needed for very high voltage X-rays. Gamma ray apparatus has the disadvantage that the source area is usually greater than the effective source area of the alternative X-ray equipment, so that photographs obtained by gamma radiography are not so sharp as X-radiographs.

# Medical applications

## Medical diagnosis

The use of X-rays in medical diagnostic work was foreshadowed in Röntgen's first paper. Within a few months of the discovery many X-ray photographs – radiographs – had been taken showing broken bones and the exact location of foreign bodies such as needles, pieces of glass and bullets embedded in patients. Damage to the joints in cases of severe arthritis, and abnormal bony growths were shown. By April 1896 several experimenters were seeking the best way of obtaining pictures of the human stomach and intestines by filling them with a suitable liquid containing X-ray opaque elements.

At first the physician employed a physicist to take radiographs for him, and some physicists set up laboratories for the purpose. The first in this country was established by A.A. Campbell Swinton in March 1896. One of Swinton's first clients was a man with a bullet in his head. The bullet was successfully located but the patient's hair began to fall out and he threatened to sue Swinton. However, the man's hair soon grew again and nothing came of the threat. The *Electrical World* of June 1896 carried the suggestion that X-rays might be used as a substitute for shaving! The possible dangers involved in the use of X-rays were not appreciated at that time. Lord Salisbury, the Prime Minister of the day, visited Swinton's laboratory and had a photograph taken of the bones of his hand. Lord Salisbury was very pleased with it and wrote to say that Lady Salisbury would like to see a photograph of the bones of her hand. Swinton duly obliged!

As soon as the practical value of radiography in medicine was established, doctors began to set up and operate their own X-ray equipment. *The Lancet*, which seems to have approached the first reports of X-rays with some scepticism, announced with astonishment on 1 February 1896 that the Belgians had decided to bring X-rays into practical use in hospitals throughout the country.

As X-ray techniques developed it became possible to improve the 'contrast' achieved in radiographs, and to distinguish organs whose X-ray opacity was only slightly different from that of the rest of the body. Tumours of the stomach and other organs were first shown radiographically in June 1896.

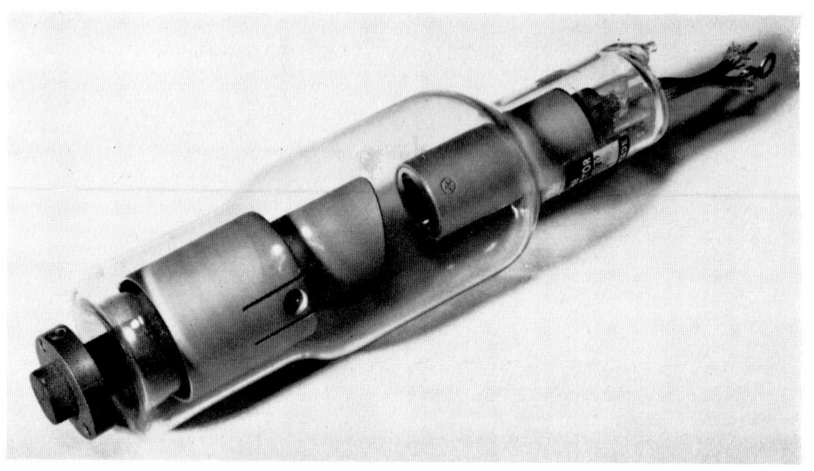

20    Line focus tube for dental use. The electron beam comes from a filament housed in the slot visible in the end of the cathode structure (right). The inclination of the target surface can be seen.

Medical radiographs were soon used as evidence in legal proceedings. An actress who fell and injured a foot in a Nottingham theatre brought an action for damages which was heard in March 1896. She produced X-ray photographs of both feet which convinced the judge and jury that a bone was out of place in the injured foot and that real injury had been sustained. In a number of similar cases Counsel argued at length about the admissibility in evidence of X-ray photographs, but they quickly gained acceptance by the courts.

Dental X-ray pictures were taken at least as early as April 1896, but X-rays were not generally applied in dentistry until about 1916. X-ray examination can reveal impacted teeth and caries which are otherwise not detectable, but because the details which need to be seen are so small, an X-ray tube with a very small X-ray source area is essential. Also the tube has to be very close to the patient's head, with consequent risk of severe electric shock. The first equipment designed specially for dental radiography appeared about 1923. This had an exposed high voltage wire from the transformer to the X-ray tube so that the patient was not entirely safe from the risk of shock. The next development was the Philips 'Metalix' set in which the tube and high voltage connection to the transformer were totally enclosed. The modern dental unit in which X-ray tube and high voltage transformer are both contained in a single, comparatively small, housing was introduced in 1933.

The modern diagnostic X-ray equipment uses a rotating anode tube which may have a power rating of fifty kilowatts or more, but the exposure times required are only a small fraction of a second. This may be contrasted with the long exposures, sometimes half an hour or more, required in the early months of radiography.

## Therapeutic X-rays

The early pioneers of X-rays had no reason to expect that the new rays would have any physiological effect, and consequently there was no reason for protecting themselves from the rays. Röntgen himself made all his experiments in a large zinc box with the tube outside, but the reason he adopted this arrangement was probably that he wanted to obtain a clearly defined beam of X-rays with no stray radiation to fog his photographic plates. However, these experimental precautions must have protected him from the rays. A few other workers protected themselves in this way. Some may have felt intuitively that rays of such penetrating power might be dangerous, but many saw no need for caution and received severe burns. Edison was one of the first to notice physiological effects from X-rays: he experienced severe pain in his eyes after several hours work.

Mention has already been made of loss of hair after radiography. Reports of more serious skin damage after exposure to X-rays appeared from time to time during 1896, and a writer in *The Lancet* of October that year described the effect as being like sunburn, from which he concluded that X-rays must be present in sunlight! During the year some investigators studied the effect of X-rays on bacteria and most reported that they could find no effect at all, although some claimed to have observed a germicidal effect.

Probably the first successful therapeutic use of X-rays was the removal of a hairy birthmark on a child by Dr L. Freund of Vienna, towards the end of 1896. Several physicians attempted to treat tuberculosis by X-rays through

21   Rotating anode tube as used in medical radiography. The anode is the central disc and its inclined edge forms the target area. The cathode is to the right of the anode and just above its centre line. The motor driving the anode is on its left. The sockets at each end are for the high voltage connections.

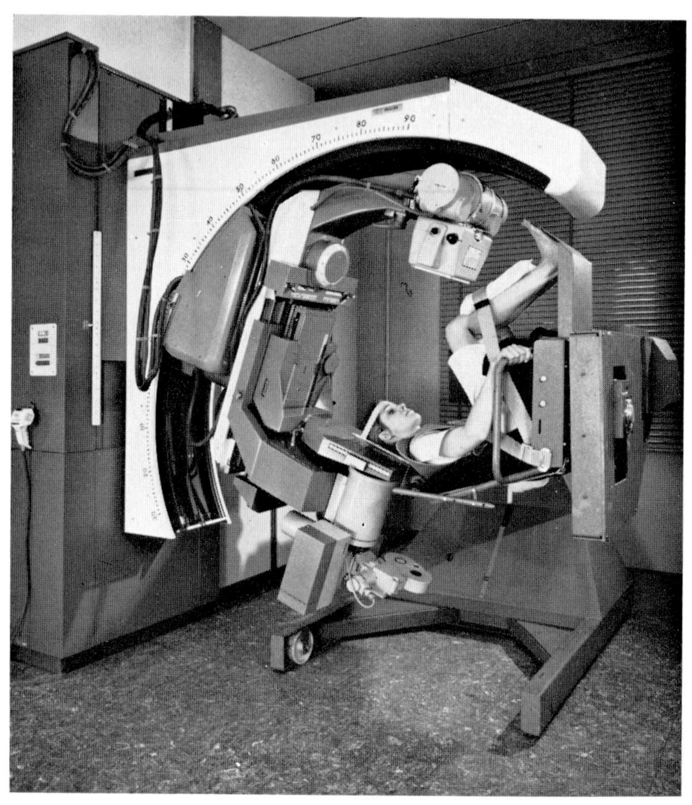

22  Modern diagnostic X-ray equipment made by Philips, enabling both the patient and the tube to be moved to any position

the lungs and a number of successes were claimed, though it is doubtful whether rays of the low intensity attainable in 1896 could have achieved the results claimed. The treatment of cancer by radiation quickly led to encouraging results. It was found that tumours were decreased in size and pain diminished, and a French physician suggested that X-rays could be a better anaesthetic than morphine.

Despite the early appreciation of the beneficial effects of X-rays on malignant tumours, it is only since the Second World War that radiotherapy has been firmly established as a distinct clinical speciality. Initially, treatment was of necessity an art rather than an exact science. The basic guide to dose was the reaction of the patient's skin. It took many years' experience after this to define the value of radiotherapy in the various types of cancers. Even today, the radiosensitivity of some rare tumours is uncertain. The growth of therapeutic X-ray technology, including the development of very high voltage X-rays, has given rise to the profession of the medical physicist. The physician is no longer pre-occupied with the apparatus and can concentrate on the patient. This is an interesting reminder of the situation in the early days of X-rays when the physicist provided and operated the equipment and the physician took his patient to the physicist for radiography.

Today surgery and radiotherapy are the major curative treatments for cancer, though some drugs are also useful. Radiotherapy also has an important role in the relief of painful and distressing symptoms occurring in the course of some common cancers.

An unfortunate and frequent sequel to radiotherapy used to be breakdown of the skin. This is seldom seen today unless the skin itself has to be treated, since the very penetrating rays used pass through the skin without affecting it. With X-rays generated above two megavolts or the gamma rays from cobalt-60 the maximum dose is absorbed 0·3 to 2 cm. beneath the skin. The limit to the total dose of radiation that is given to a tumour is determined by the tolerance of the adjacent normal tissues since these cannot be spared in the way other areas of tissue are, by using multiple X-ray beams intersecting at the place it is desired to treat. The damage done by X-rays is not selective for malignant cells. A proportion of both normal and malignant cells are killed but normal tissues are reconstituted by the orderly division of a population of reserve cells which is not present in the tumour. The chief factor determining normal tissue tolerance appears to be irreversible damage to the blood vessels which progressively worsens as time passes.

Until improved drugs are developed, radiotherapy will retain its leading role in the treatment of cancer.


# Non–medical applications

### Industrial applications of X-rays

The possibility of using X-rays for industrial testing was realised from the start. In his first publication Röntgen had drawn attention to the fact that with the new rays it was possible to detect voids in an apparently solid piece of metal. Welded joints could be subjected to an examination which would show clearly whether the parts welded together were really fused into a single piece of metal throughout their whole thickness, and not merely joined by a surface layer. Internal cracks and unwanted inclusions of foreign matter in castings could readily be detected.

However, although the possible industrial uses of X-rays were appreciated as early as the medical possibilities, the actual introduction of industrial X-ray equipment was rather slower. This was because in most cases industrial applications require X-rays of far higher penetrating power than the medical. X-rays of sufficient penetration to radiograph the human body can only penetrate about one-tenth of an inch of steel. X-ray equipment adequate for medical diagnostic work could be assembled from components (Crookes' tube, induction coil and fluorescent screen or photographic plates) which were available 'off the shelf' in 1896. Industry needed more advanced X-ray tubes requiring improved power supplies. In 1922 the greatest thickness of steel which could be examined by X-rays was about six inches. At the time of writing (1969) the limit is about twenty inches.

One of the earliest industrial users of X-rays was the British Post Office. About 1900 they installed equipment for examining guttapercha, which was used for the insulation of submarine telegraph cables, and later they used X-rays to inspect the lead sheaths of underground cables. Another early industrial use of X-rays was examining live oysters for pearls. The 1914–1918 war stimulated the development of X-ray testing. Shells could be examined to see that the metal itself was sound and that it was properly filled. Because of the importance of structural woodwork in the aeroplane industry the Aeronautical Inspection Department began an investigation in 1918 into the value of X-rays for inspecting timber.

One of the most recent applications is the checking of welds joining sections of pipe in the National Gas Board's grid. The pipes, which vary from 20 to 40 inches in diameter and weigh several tons each, are welded *in situ* and every weld is radiographed. For this work a battery operated 'X-ray crawler' is employed which travels inside the pipe. The crawler carries an X-ray tube which has a specially shaped target so as to produce a ring of radiation. The operator (outside the pipe) fixes a strip of X-ray sensitive paper around the tube over the welded joint, and also places an isotope on the pipe at a predetermined position in relation to the weld. The crawler moves forward under its own power until it detects the isotope. The X-ray tube is energised for the appropriate time and the crawler then automatically moves forward to the next isotope 'signal' on the outside of the pipe. The operator never has to be near the crawler when it is generating X-rays, and the length of pipe which can be X-rayed without the crawler requiring any attention is limited only by the capacity of the batteries.

For routine inspection of mass produced items it is sometimes possible to X-ray several components simultaneously, with a consequent saving in cost of photographic film. For example, a large number of semi-conductor devices may be laid in a plastic jig and details of their internal construction checked on a single radiograph.

An interesting recent application of X-rays is in potato harvesting machinery which distinguishes potatoes from stones or clods of earth by using the fact that potatoes are less opaque to X-rays than either stones or earth. The potatoes and other material picked up by the harvester are passed through a beam of X-rays and the mechanism deflects the potatoes on to one conveyor belt and the rubbish on to another.

## X-rays in the study of paintings

X-ray examination of a painting can be of great assistance to both the historian of art and the conservationist. Some of the pigments used in painting contain metals of high atomic weight (usually lead or mercury) and these pigments are consequently opaque to X-rays. A radiograph of a painting will reveal the use made by the artist of such pigments, even though another colour may have been superimposed. By comparing an ordinary photograph with an X-ray photograph of the same painting it is possible to learn more about the artist's technique, and sometimes to reveal alterations which have been made. Such alterations may have been made by the artist himself before the work was finished, or they may be later alterations, often done for religious or moral reasons. Once the original form has been determined it can be restored if desired.

The most important pigments from the point of view of X-ray examination are the white lead oxide (PbO$_2$) and the vermilion red mercuric sulphide (HgS). The illustrations on the following pages (figures 23a 23b and 24a 24b) are black and white photographs and X-radiographs of paintings by Antonello and Rembrandt. The radiographs illustrate the difference in technique of the two artists. Antonello painted on a white ground using pigments comparatively transparent to both visible light and X-rays. Rembrandt painted on a dark ground and built up the modelling of his flesh paint with thick lead white.

The radiograph of Antonello's portrait reveals a number of interesting features. The forehead has been damaged at some time and repaired with a paint which is a perfect match for colour, but the repair is easily seen in the radiograph because the restorer used a paint of greater X-ray opacity than the pigment used by Antonello himself. The painting is on a wooden panel whose grain runs horizontally, which is unusual in a painting whose height is greater than its width. There are worm tunnels, dark lines to the left of the chin, but the wood is in reasonable condition and not cracked. There is an artist's alteration or 'pentimento' in the position of the eyes which as first painted were turned less far to the right than we now see them. There is no suggestion that the eyes as we now see them are painted not by the artist but by a later hand, for the style and technique are consistent with the rest of the picture.

The Rembrandt is on canvas, and in the original radiograph the weave of the canvas can be seen. The lighter horizontal band across the top of the radiograph represents the wood of the stretcher on which the canvas is stretched. The nails fixing the canvas to it, being of solid metal, show up with great clarity on the radiograph. At the top left corner of the radiograph can be seen a knot in the wood of the stretcher.

Because the layers of pigment being examined are very thin, X-rays of low penetration ('soft' X-rays) are required in order that sufficient contrast may be obtained. Tube voltages of about 10kV are typical, and X-ray tubes with a beryllium window are sometimes used, since beryllium is more transparent to soft X-rays than ordinary glass.

The study of paintings is yet another application of Röntgen's discovery which was taken up within the first few months of 1896. In his original paper Röntgen records that a coat of lead paint on a piece of wood is fairly opaque to X-rays, whereas the wood itself is transparent. A number of paintings were examined by German workers, including one by Albrecht Dürer whose authenticity was doubted. The radiograph revealed Dürer's signature even though it was so faded as to be invisible to the unassisted eye. The London *Electrical Review* reporting this in 1897 made the prophecy that 'the application of roentgen rays may prove of considerable value to picture dealers and others in detecting fraudulent imitations of valuable paintings'. The German radiologist Alexander Faber obtained a German Patent in 1914 for 'A procedure for the determination of overpainting in oil paintings and similar objects by the preparation of radiographs'.

Today radiography is a well recognised and widely used tool in the scientific study of works of art, and a large number of paintings in many countries have been subjected to X-ray examination.

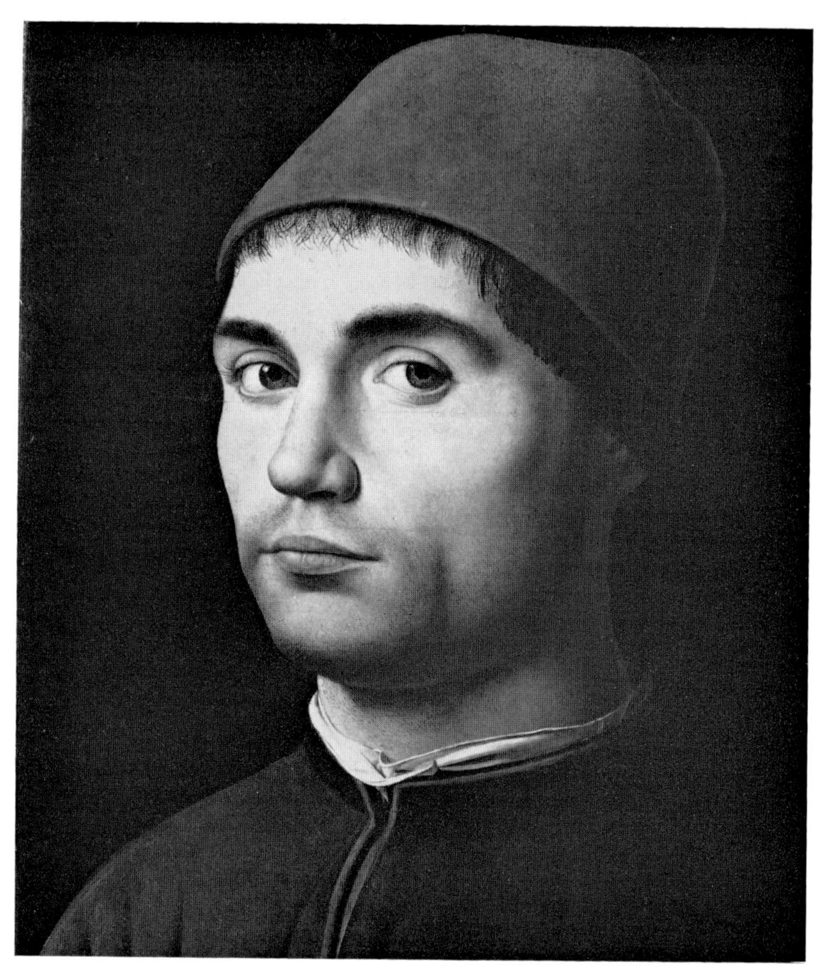

23a  Antonello da Messina: Portrait of a man. (National Gallery)

**23b**  X-ray photograph of **a**

**24a**  Rembrandt: Portrait of the Painter in Old Age. (National Gallery)

24**b**  X-ray photograph of **a**

# The nature and measurement of X–rays

X-rays are an electromagnetic radiation with wavelengths in the range $10^{-14}$ to $10^{-8}$ metres approximately. This is much shorter than the wavelength of visible light (about 3 to $7 \times 10^{-7}$ metres) or the wavelengths used in radio-communication ($10^{-1}$ to $10^4$ metres). When preparing his first paper on the new rays Röntgen thought that they must be fundamentally different from visible light because he could not reflect or diffract them. It was eventually realised that the wavelength of X-rays must be very short. Reflection requires a mirror whose surface roughness is small compared with the wavelength of the rays being reflected, and no mirror was smooth enough to reflect X-rays. Similarly no man-made diffraction grating was fine enough to diffract X-rays.

In 1912 the physicist Max von Laue was working on the structure of crystals, and it occurred to him that the atomic lattice of a crystal might form a natural diffraction grating which could diffract X-rays. His theory proved to be correct, and as well as revealing something of the fundamental nature of X-rays he formed the basis of the science of X-ray Crystallography – the analysis of crystal structures and dimensions by the study of X-ray diffraction patterns. Also in 1912 Sir Lawrence Bragg reflected X-rays from a cleavage face of mica.

The measurement of X-rays is a complex subject. There are two principal characteristics to be defined: the *penetration* (or *penetrating power*, or *quality*) and the *intensity*. The penetration can be defined by the wavelength of the radiation; in practice it is defined by the kilovoltage applied to the X-ray tube. The terms 'hard' and 'soft' refer to the penetration of the rays, but they are purely comparative terms with no precise meaning.

The intensity is the measure of the energy in the X-rays. One of the first units of intensity was the H unit, defined by Holzknecht as one-third of 'the quantity of rays compatible with the integrity of the skin, and only causing on the face of an adult a very slight inflammatory action'. The usual method of measuring intensity accurately is to employ an ionisation chamber and determine electrically the ionisation produced in air by the passage of the X-rays being monitored.

The unit of intensity of X-rays is the röntgen, defined as the quantity of radiation 'such that the associated corpuscular emission per 0·001293 gram of air [1 cc at 0°C] produces, in air, ions carrying 1 electrostatic unit of quantity of electricity of either sign'. The absorbed dosage is important to the medical radiologist since biological effects depend on the energy absorbed by tissue, and in diagnostic radiography the total absorbed dosage is kept as small as possible.

The Rad is the unit of absorbed dose of ionising radiation and corresponds to an energy absorption of 100 ergs per gram of irradiated material.

In practical radiographic examinations, whether medical or industrial, it is common practice to specify two quantities: the kilovoltage on the tube and the product of the tube current (which determines the intensity) and the duration expressed in milliamp-seconds.

Printed in England for Her Majesty's Stationery Office by Swindon Press Ltd., Swindon, Wilts.
Dd. 501681   K60   Gp 469